KB263225

일러스트 프로덕트비젼

출판사 : 하랑출판
주　소 : 서울시 중구 퇴계로28길 8
전　화 : 02-2263-3337

이 책의 저작권은 하랑출판에 있습니다.

일러스트 프로덕트비젼

발행일 : 2025년 08월 24일
출판사 : 하랑출판
주　소 : 서울시 중구 퇴계로28길 8
전　화 : 02-2263-3337

이 책의 저작권은 하랑출판에 있습니다.

Division

포토디자인과 타이포그래픽　006
다양한 광고를 위한 애니메이션　018

컴퍼니 일러스트　083
포토그래픽을 위한 그랜드 플랜　096

복고풍의 유럽 애드버타이징　127
인물표현의 고전적인 해석　131

자연주의 일러스트와 현실주의 일러스트　143
핸드드로잉 기법의 고전 작품　154

PHOTO Hans Starck   GRAPHIC Raphael Milczarek   CORPORATION JWT, New York   COPY Humberto Jiron

GRAPHIC Timsy Vadhani   CORPORATION Euro RSCG, Dubai

GRAPHIC Ary Nogueira

PHOTO Alexandre Catan   GRAPHIC Marcos Almirante   CORPORATION AlmapBBDO, São Paulo   COPY Marcos Almirante

PHOTO Oliver Gast   GRAPHIC Tolga Büyükdoganay, Sebastian Kaufmann   CORPORATION Demner, Merlicek & Bergmann, Vienna

PHOTO Anthony Redpath   GRAPHIC Colin Hart

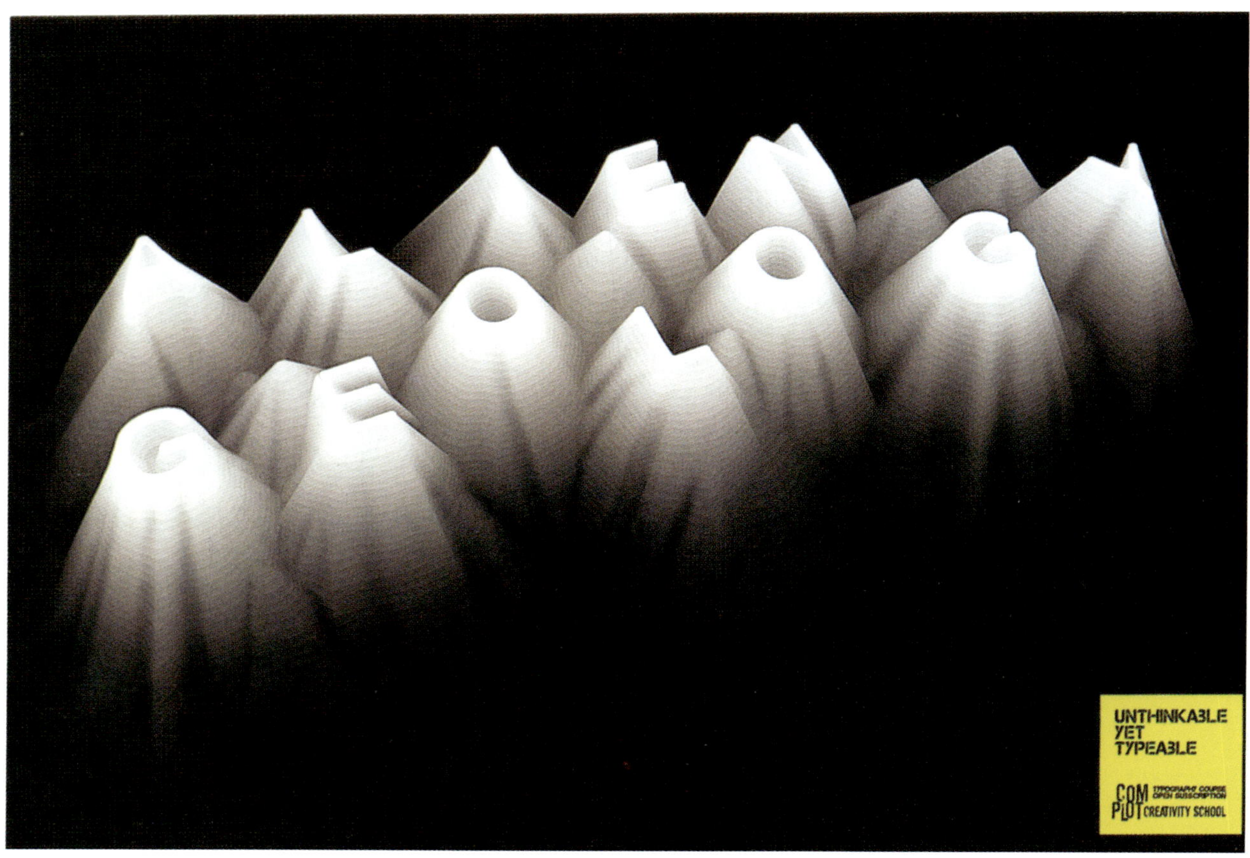

GRAPHIC Lucas Jatobá   CORPORATION Contrapunto, Barcelona   COPY Emma Piquer

GRAPHIC Andreas Junus   CORPORATION Bates141, Jakarta   COPY Iyan Susanto

PHOTO Metz & Racine, John Akehurts   GRAPHIC Emer Stamp   CORPORATION Adam&Eve, London   COPY Ben Tollett

PHOTO Zena Holloway, Lorentz Gullachsen   GRAPHIC Fabrizio Pozza   CORPORATION JWT, Milan   COPY Francesco Muzzopappa

PHOTO Patrick Frischknecht GRAPHIC Dan Strasser, Alexander Reiss CORPORATION DDB, Dusseldorf, Germany

PHOTO Lars Topelmann   GRAPHIC Brent Anderson, Clark Evans   CORPORATION TBWA\Chiat\Day, Los Angeles

GRAPHIC Bianco Ramirez

GRAPHIC Juan Posada   CORPORATION Ogilvy & Mather, Bogotá   COPY Diego Ortiz, John Forero

PHOTO Bob Clarke   GRAPHIC Mike Cozens   PHOTO John Hegarty   CORPORATION BBH, London

GRAPHIC Graham Watson   CORPORATION BBH, London   COPY Chris Herring

PHOTO Tierney Gearon   GRAPHIC Pascal Midavaine   CORPORATION BDDP, Paris   COPY Pascale Chadenat

PHOTO Tom Cwenar　　GRAPHIC Derek Julin　　CORPORATION Brunner, Pittsburgh　　COPY Eric Schlauch

PHOTO Geison Genga   GRAPHIC Ricardo Big Passos   CORPORATION DDB, São Paulo   COPY Antero Neto

GRAPHIC Delphine Arnol

PHOTO Yann Le Pape   GRAPHIC Benoît Blumberger   CORPORATION Publicis, Paris

PHOTO Davide Bodini   GRAPHIC Gabriele di Donato

PHOTO Ferdinando Galletti   GRAPHIC Ferdinando Galletti   CORPORATION Lowe Pirella, Milan COPY Atonino Munafò

PHOTO Andrew Chapman   GRAPHIC Thomas Dooley, Matt Leavitt   CORPORATION TDA Advertising & Design, Boulder

PHOTO Mats Cordt   GRAPHIC Christian Sommer, Ivo Hlavac   CORPORATION Serviceplan, Munich

PHOTO Yann Le Pape   GRAPHIC Guillaume Auboyneau   CORPORATION Y&R, Paris

PHOTO Justen Lacoursiere   GRAPHIC Kelsey Horne   CORPORATION TAXI, Calgary   COPY Nick Asik

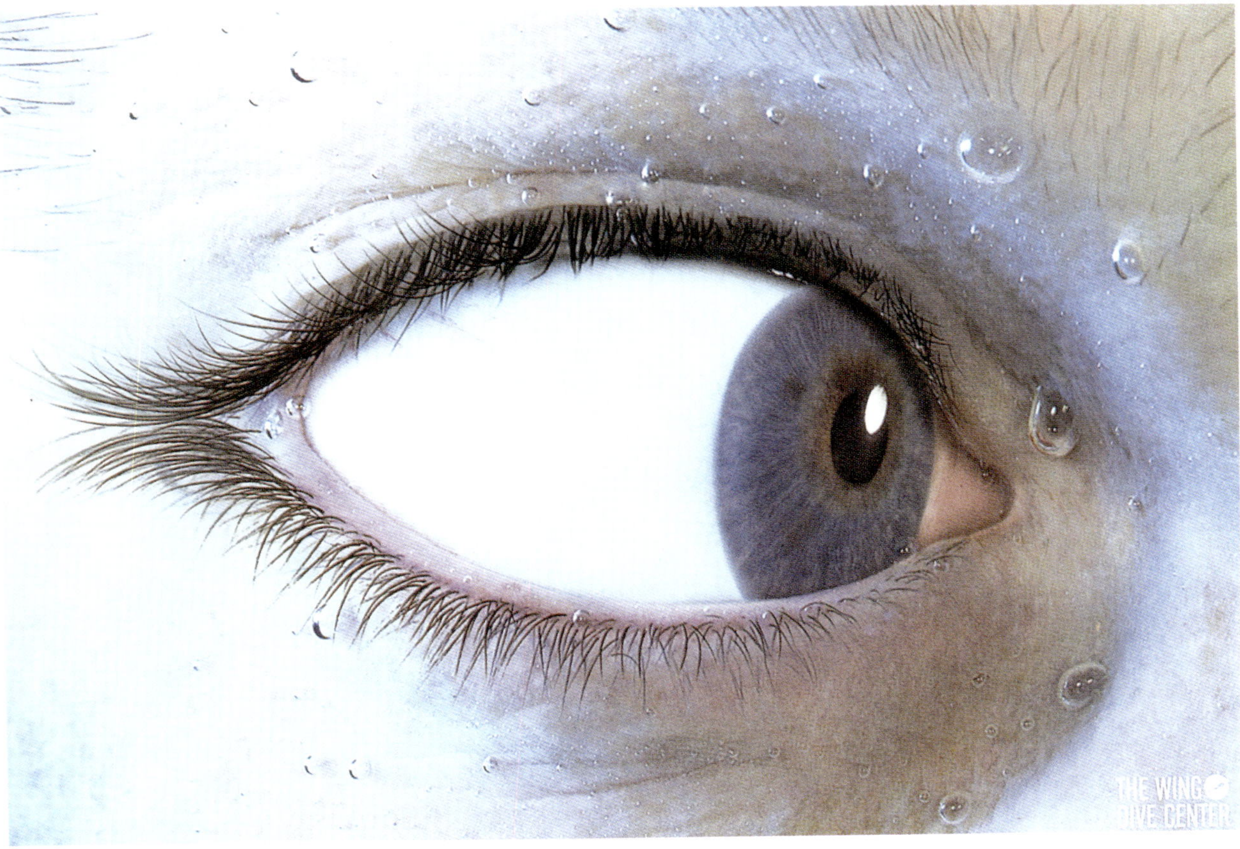

PHOTO Chup Cheevit   GRAPHIC Sirichai Thanachoksombat   CORPORATION Publicis, Bangkok

GRAPHIC Mikkel Kroijer, Anders Moller   CORPORATION Grey, Copenha

PHOTO István Lábady   GRAPHIC Balázs Kismarty-Lechner   CORPORATION Ogilvy & Mather, Budapest

PHOTO Fernando Ziviani   GRAPHIC Gean Santos   CORPORATION Wow!, Curitiba, Brazil

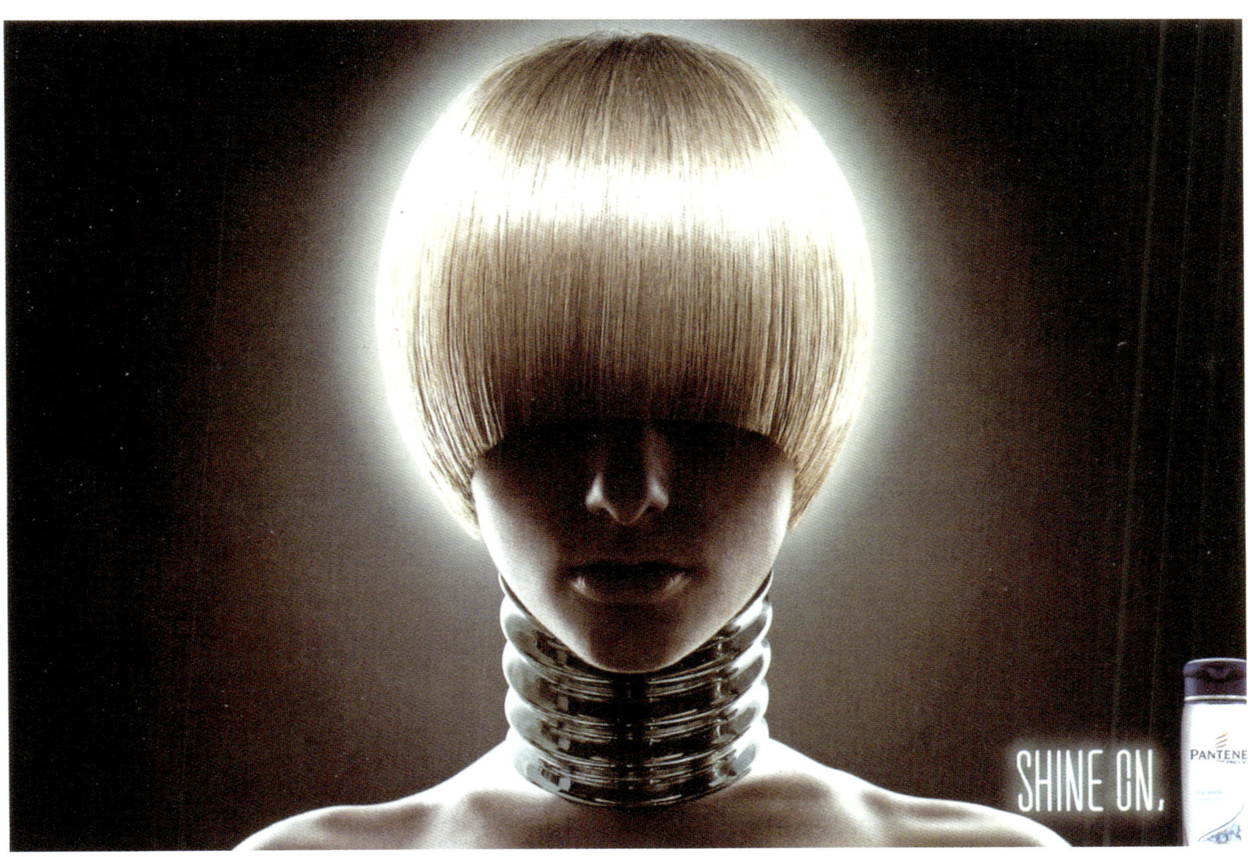

PHOTO Connie Hong   GRAPHIC Doris Leung   CORPORATION Grey, Hong Kong

PHOTO David Prior GRAPHIC Alexa Craner CORPORATION Ogilvy, Johannesburg COPY David Krueger

PHOTO Liu Zhu   GRAPHIC Ning Zhou, Nils Andersson, Jacky Lung   CORPORATION Ogilvy, Beijing   COPY Duan Qing

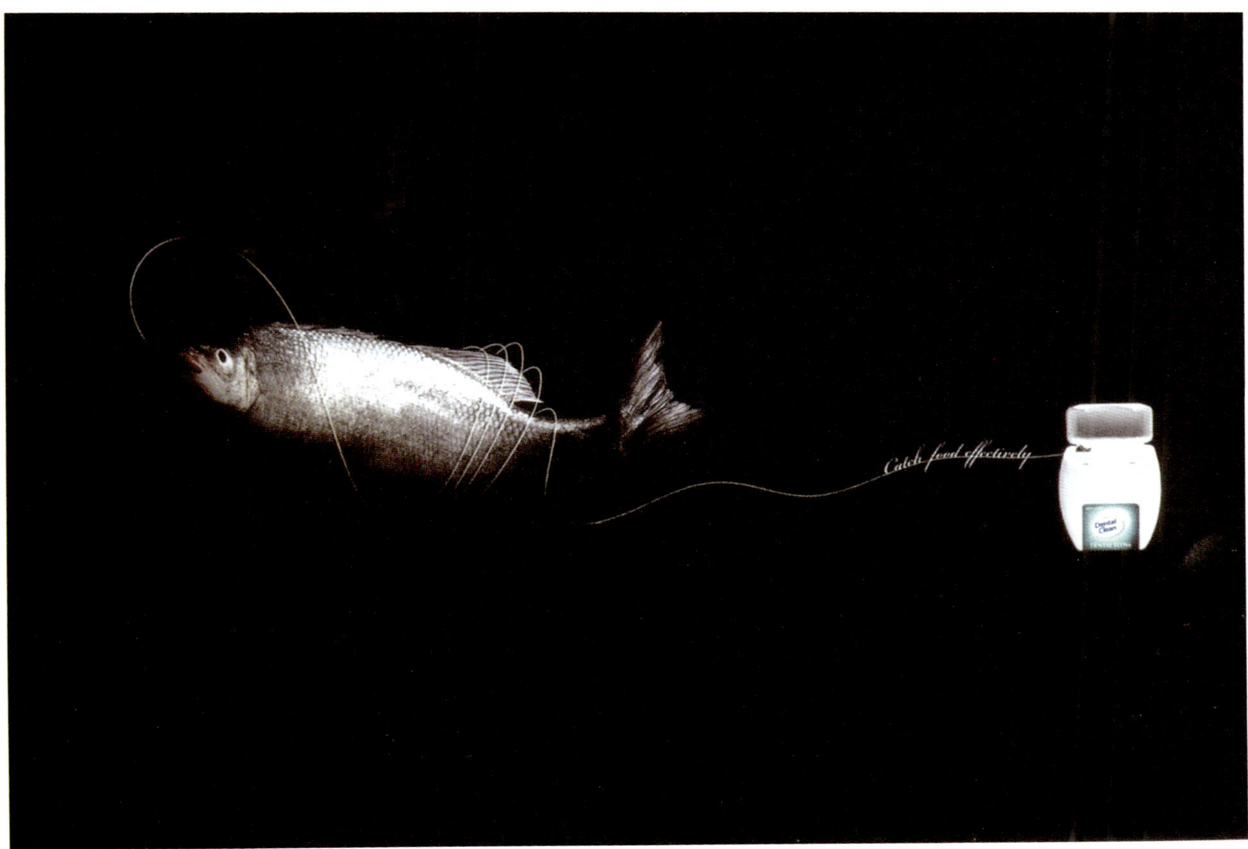

PHOTO Steve Kosman   GRAPHIC Ahmad Beck   CORPORATION Impact BBDO, Riadh   COPY Ahmad Beck

PHOTO Ricardo Salaman GRAPHIC Miguel Angel Cerdeira, Patricio Céspedes CORPORATION Grey, Santiago

GRAPHIC Christopher Bowsher   CORPORATION Dye Holloway Murray, London   COPY Frances Leach

PHOTO Tim Brett Day   GRAPHIC Steve Williams CORPORATION RKCR/Y&R, London

PHOTO Marc Wuchner   GRAPHIC Khai Doan, Kathrin Waschek   CORPORATION JWT, Frankfurt am Main   COPY Jan Koehler

PHOTO Jacek Kołodziejski   GRAPHIC Saatchi & Saatchi, Warsaw   CORPORATION Magdalena Nowakowska, Johan Pasternak

PHOTO Hans Starck   GRAPHIC Raphael Milczarek   CORPORATION JWT, New York   COPY Humberto Jiron

ENO
Explore the world

ENO
Explore the world

PHOTO Erwin Olaf   GRAPHIC Alexandre Lagoët   CORPORATION Saatchi & Saatchi, Amsterdam   COPY Rick Coolegem

PHOTO Pradeep Dasgu GRAPHIC Sudeepa Ghosh CORPORATION Ogilvy & Mather, Gurgaon, India COPY Mayur Hola

PHOTO Kimmo Virtanen GRAPHIC Tuukka Tujula CORPORATION TBWA\PHS, Helsinki COPY Taro Korhonen

PHOTO Ed McCulloch   GRAPHIC Mark Rawlins, Brandon Knowlden   CORPORATION Struck, Salt Lake City

GRAPHIC Connie Lo, Pranussadej Tantipong, Martin Tong Sze Lok   CORPORATION Leo Burnett, Shanghai

GRAPHIC Dave Androliakos   CORPORATION DaVinciEdisonBell, Johannesburg   COPY Angela Collins, Patrick Robertson

GRAPHIC Gustavo Victorino CORPORATION DDB, São Paulo COPY Otavio Schiavon

GRAPHIC Sthefan Ko   CORPORATION JWT, São Paulo   COPY Luiz Filipin

GRAPHIC Ryan Anderson   CORPORATION Richter7, Salt Lake City

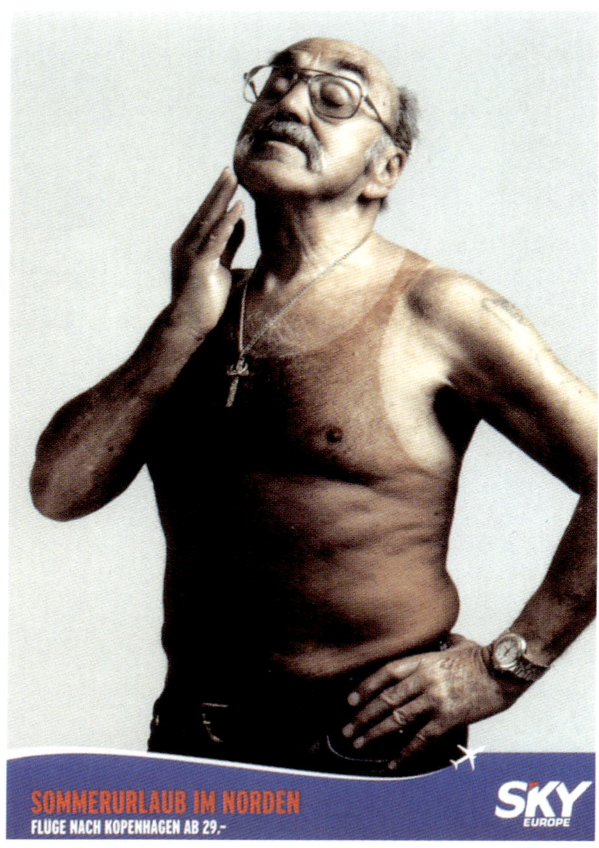

PHOTO Bernd Preiml   GRAPHIC Andreas Gesierich

PHOTO Esteban Sosnitsky   GRAPHIC Rodrigo Tarquino   CORPORATION Leo Burnett, Bogotá   COPY Mauricio Rocha

PHOTO Markku Lahdesmaki GRAPHIC Sean Ohlenkamp CORPORATION M&C Saatchi, Los Angeles COPY Maria Smith

PHOTO John Sylvester

PHOTO Paul Baglole

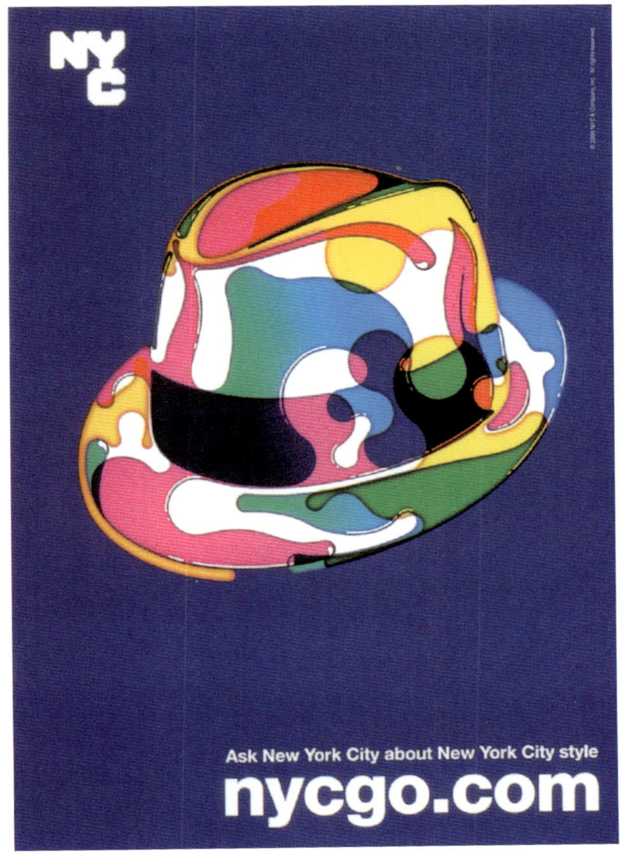

GRAPHIC Nick Klinkert, Pelle Sjoenell

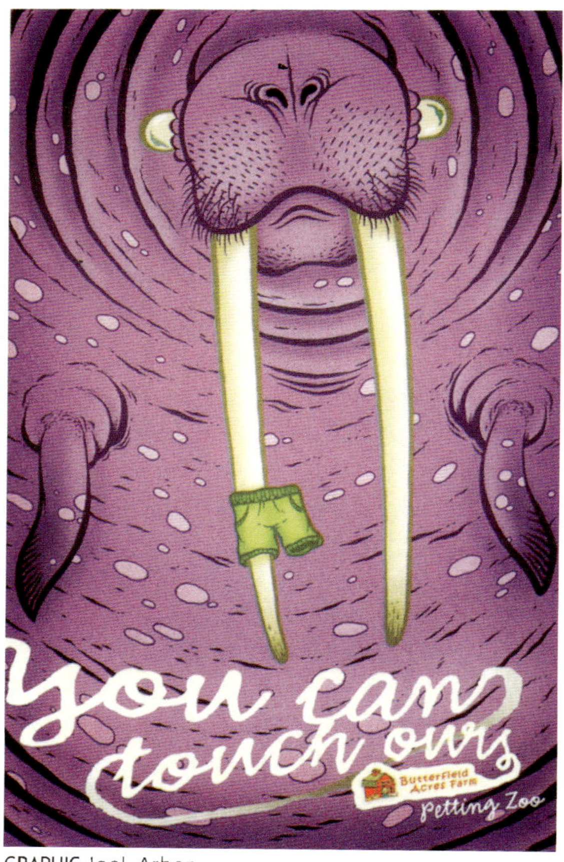

GRAPHIC Joel Arbez

GRAPHIC Anne Taylor   CORPORATION Preston Kelly, Minneapolis   COPY Terry Thomas I

PHOTO Thomas Mailaender   GRAPHIC Erik Kessels   CORPORATION KesselsKramer, Amsterdam

GRAPHIC Yuval Davidov, Gal Shkedi **CORPORATION** Gitam BBDO, Tel Aviv

PHOTO Jean-Yves Lemoigne GRAPHIC Thierry Chiumino, Marie Farge, Eve Roussou CORPORATION Ogilvy & Mather, Paris

PHOTO Miles Aldridge   GRAPHIC Michele Mariani, Andrea Lantelme   CORPORATION Armando Testa, Turin   COPY Michele Pieri

GRAPHIC Jörg Dittmann   CORPORATION Kolle Rebbe, Hamburg   COPY Stefan Wübbe

GRAPHIC Thomas Knüwer   COPY Florian Ludwig

PHOTO Josefina   GRAPHIC Jorge Muñoz   CORPORATION Prolam/Y&R, Santiago de Chile

PHOTO Lúcio Cunha   GRAPHIC Keka Morelle   CORPORATION F/Nazca Saatchi & Saatchi, São Paulo   COPY André Faria

GRAPHIC Diana Sukopp, Volker Schrader   CORPORATION Publicis, Frankfurt am Main   COPY Stephan Ganser

PHOTO Lado Alexi   GRAPHIC Vanessa Rabea Schrooten, Felix Taubert, Dirk Haeusermann   CORPORATION Jung von Matt, Hamburg

GRAPHIC Diya Ajit, Vincen Raffray

GRAPHIC Erik Heisholt, Martin Holm   CORPORATION TBWA, Oslo   COPY Erik Heisholt

GRAPHIC Mother, London   CORPORATION Mother, London   COPY Mother, London

GRAPHIC Tilman Gossner CORPORATION Jung von Matt,

is it in you?
Gatorade
7 KILOMETERS
BEHIND YOU
BREATHE IN BREATHE OUT
AIR WON'T RUN OUT ON YOU

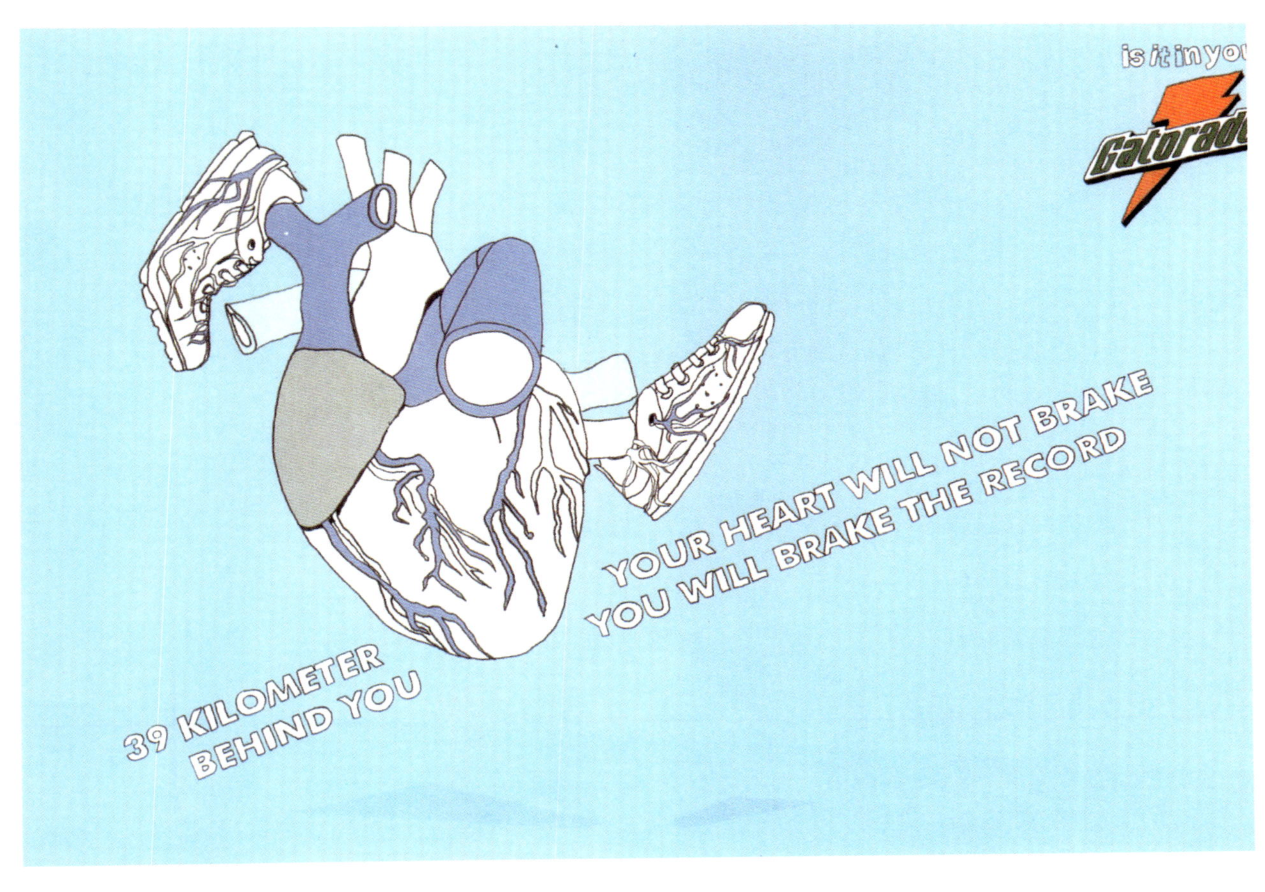

is it in you?
Gatorade
YOUR HEART WILL NOT BRAKE
YOU WILL BRAKE THE RECORD
39 KILOMETER
BEHIND YOU

PHOTO MacLaren McCann, Toronto GRAPHIC Robert Kingston CORPORATION Wade Hesson, Nancy Crimi COPY Frank Hoedl

GRAPHIC Katheline Besseling   COPY Nerine Lombaard, Angela Bowes

GRAPHIC Toni Eisner   COPY Toni Eisner

GRAPHIC Pasha Ganin   COPY Pasha Ganin

GRAPHIC Ignácio Pérez Brea   COPY Juan Pablo Papaleo

GRAPHIC Alicia Pol   COPY Steven Hancock

GRAPHIC Alicia Pol   COPY Alicia Pol

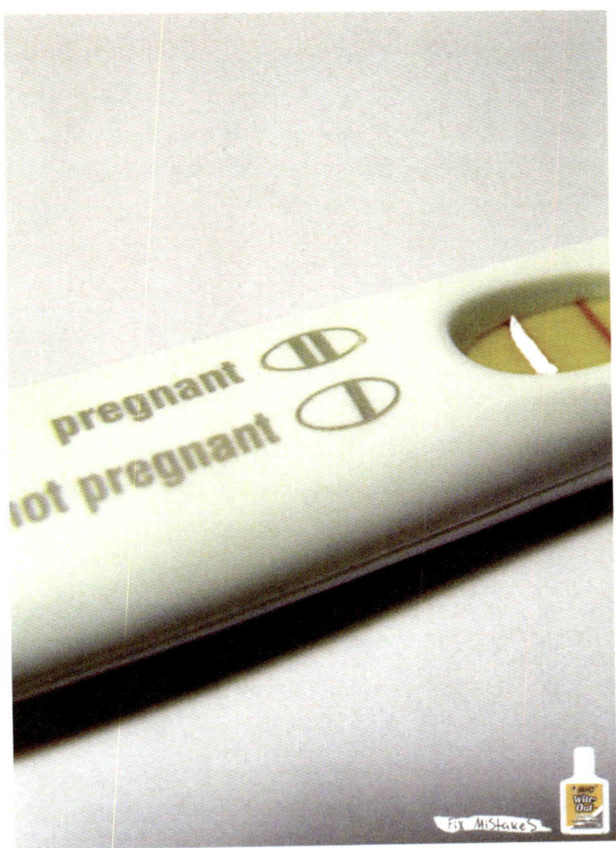

GRAPHIC Blake Romine COPY Blake Romine

GRAPHIC Linda Weitgasser   COPY Alexandra Sattlecker

PHOTO Michael Pretz GRAPHIC Michael Pretz COPY Michael Pretz

GRAPHIC Aman Gulati COPY Aron Tipton

GRAPHIC Jessa Simmons   COPY Jess Kline

GRAPHIC Nicolas Comastri COPY Jean Guérin

GRAPHIC Diego Machado COPY Ricardo

GRAPHIC Clémence Brousse   COPY Clémence Brousse

PHOTO Suresh Natarajan   GRAPHIC Ryan Menezes   CORPORATION McCann Erickson, Mumbai   COPY Ryan Menezes

PHOTO Arata Dodo GRAPHIC Kotaro Hirano

GRAPHIC Klaus Seethaler   COPY Klaus Seethaler

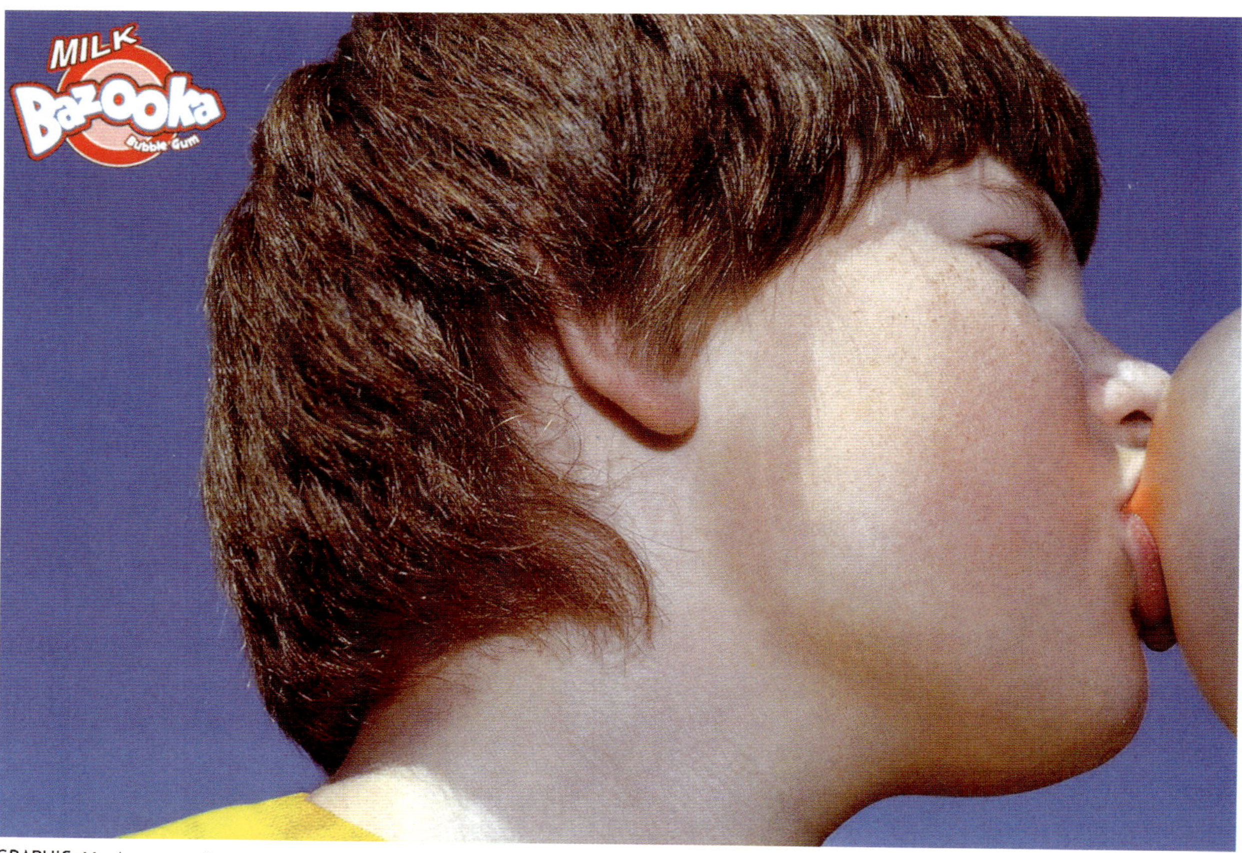

GRAPHIC Marlon von Franquemont, Reinier Gorissen   COPY Marlon von Franquemont, Reinier Gorissen

GRAPHIC Fabian Jung, Sebastian Kamp

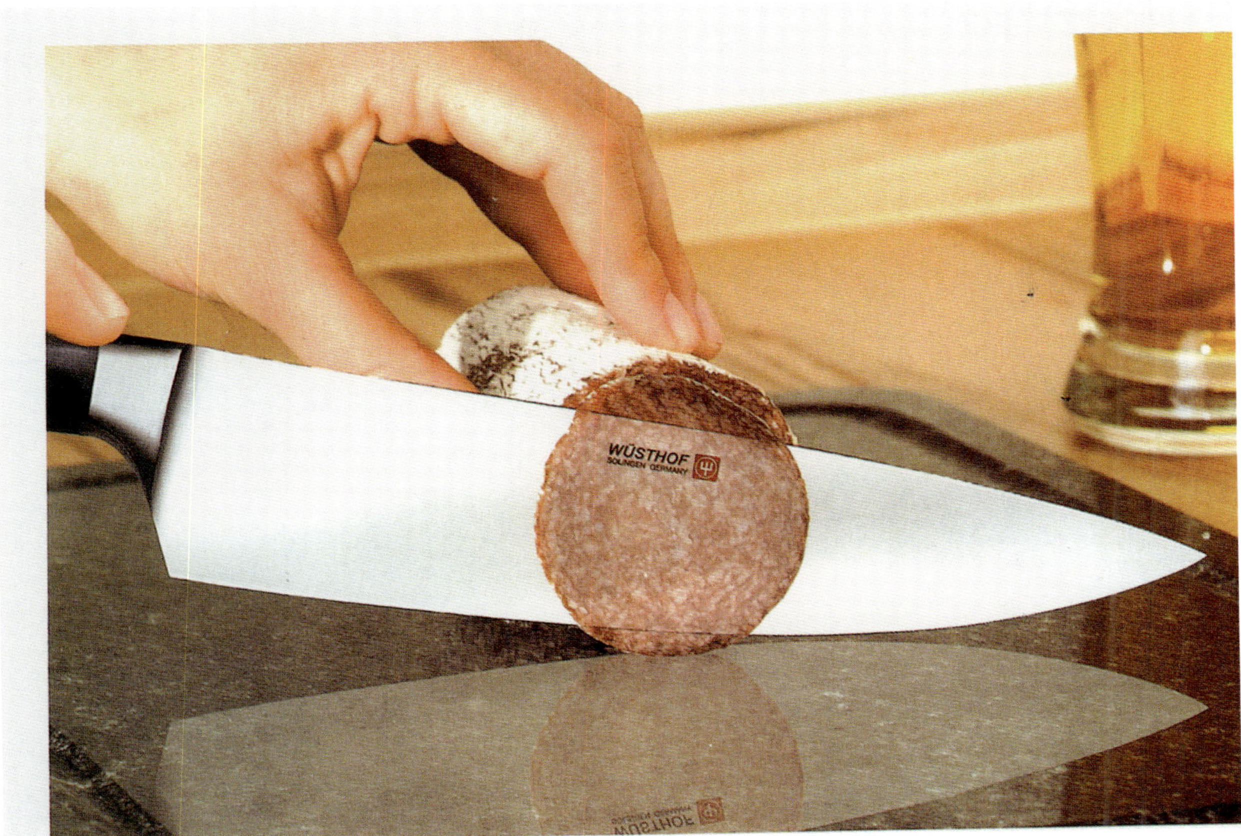

GRAPHIC Gerrit-John Gerischer, Benny Hoffmann

GRAPHIC Lisa Erwall

GRAPHIC Pasha Ganin   COPY Pasha Ganin

PHOTO Nils Sandmeier   GRAPHIC Samuel Huber   COPY Oddbjørn Stensrud

PHOTO Sebastian Kamp, Stefan Bolten

GRAPHIC Caryn Jendro   COPY Caryn Jendro

GRAPHIC Maximilian Pinegger   COPY Justin Salice-Stephan

GRAPHIC Rob Fletcher   CORPORATION Isobel, Londor

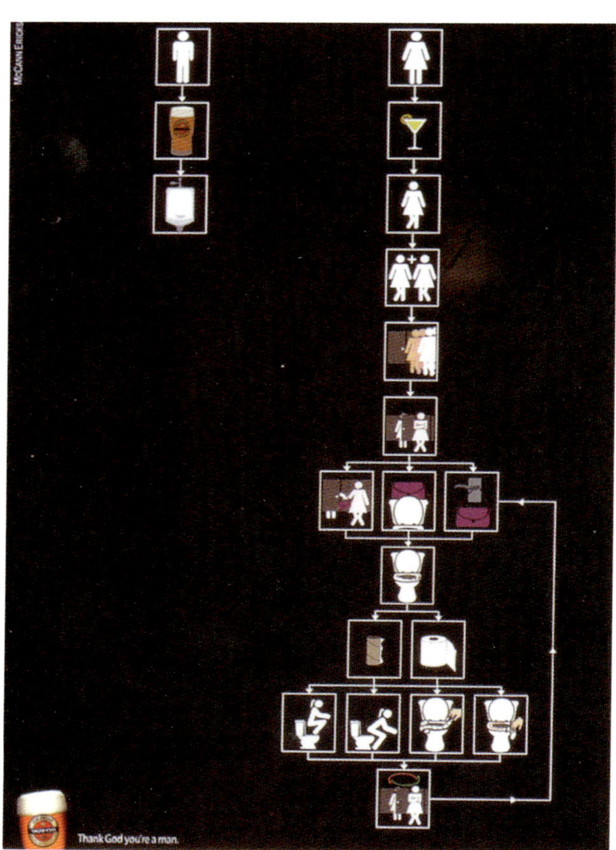

GRAPHIC Geva Gershon CORPORATION McCann Erickson, Tel Aviv

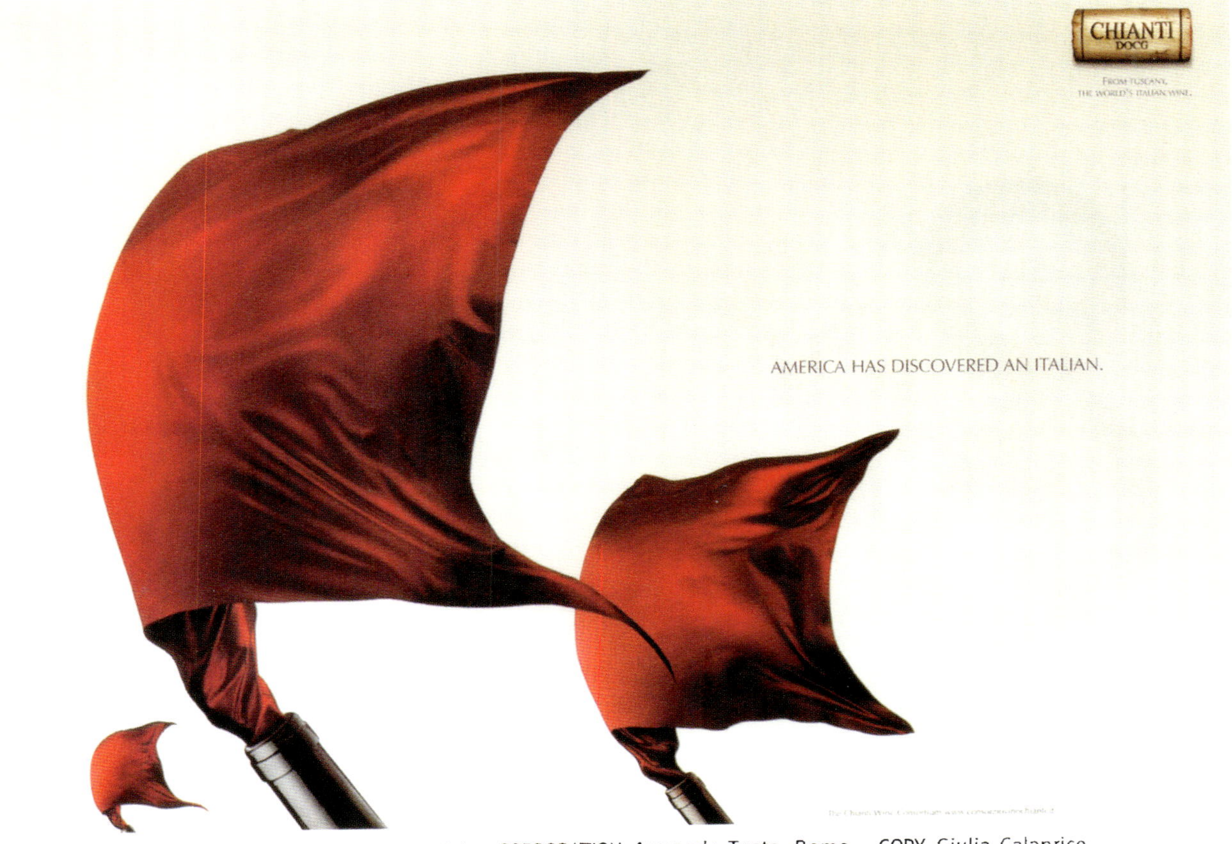

PHOTO Maurizio Cigogi   GRAPHIC Roberto Orfei   CORPORATION Armando Testa, Rome   COPY Giulia Calaprice

PHOTO Marcel van der Vlugt   GRAPHIC ..., staat

GRAPHIC Jim Haven, Steve Cullen, Kristi Flango, Chris Campbell   CORPORATION Creature, Seattle

PHOTO Christophe Gilbert   GRAPHIC Ty Baker   CORPORATION Momentum, New York

PHOTO Christophe Gilbert   GRAPHIC Alexey Ryzhof   CORPORATION Mojo, Moscow   COPY Alexey Mudrov

PHOTO Tal Silverman   GRAPHIC Phillip Meyler, Darren Keff

PHOTO Andreas Smetana   GRAPHIC Chris Northam   CORPORATION George Patterson Y&R, Melbourne   COPY Simon Bagnasco

CORPORATION Michael Hansen

PHOTO Carioca   GRAPHIC Kamila Kubiak, Michael Hansen   CORPORATION Change Communications, Warsaw

PHOTO Stephen Langdon   GRAPHIC Lisa Fedyszyn   CORPORATION Colenso BBDO, Auckland   COPY Jonathan McMahon

PHOTO Patrice Lange   GRAPHIC Oliver Brkitsch, Dirk Siebenhaar   CORPORATION Grabarz & Partner, Hamburg

GALERIA
INNO
DEPARTMENT STORE

PHOTO Christophe Gilbert   GRAPHIC Claudine Mergaerts

PHOTO Manolo Moran   GRAPHIC Max Geraldo   CORPORATION DDB, São Paulo   COPY Aricio Fortes

PHOTO Nilesh Patankar   GRAPHIC Uday Parkar, Vijayan Acharya

PHOTO Steve Koh   GRAPHIC Zaidi Awang   CORPORATION Leo Burnett, Kuala Lumpur   COPY Jovian Lee, Zaidi Awang

PHOTO Jean-Noël Leblanc Bontemps   GRAPHIC Jean-Luc Collard   CORPORATION Euro RSCG 360, Paris   COPY Valerie Dietrich

PHOTO Avadhut Hembade   GRAPHIC Manasi Kotian, Heeral Desai   CORPORATION Ogilvy & Mather, Mumbai

PHOTO Rick Guest   GRAPHIC Graeme Hall, Hunter Somerville   CORPORATION DDB, London   COPY Graeme Hall, Hunter Somerville

WE DON'T THINK GREED,
DISHONESTY OR EXPLOITATION
MAKE YOUR SKIN ANY SMOOTHER.
We've always strived to bring you beauty products bursting with effectiveness. And we believe that people have the right to a fair wage and to be treated with respect. We work directly with over 30 Community Trade suppliers to build businesses in more than 20 countries helping 25,000 people earn a sustainable income. We don't do this because it's fashionable; we do it because we believe it's the only way to do business.
THE BODY SHOP

THERE'S SOMETHING MORE
ALARMING THAN
HIV, DOMESTIC VIOLENCE
AND EXPLOITATION.
IGNORING THEM.
THE BODY SHOP

PHOTO Manolo Moran   GRAPHIC Marcelo Rizério   CORPORATION AlmapBBDO, São Paulo   COPY Marcos Almirante

PHOTO Philip Rostron GRAPHIC Basil Douglas Cowieson CORPORATION Saatchi & Saatchi, Toronto COPY Tal Wagman

GRAPHIC Francois Boshoff, Glenn Jeffery   CORPORATION Volcano, Bryanston, South Africa   COPY Francois Boshoff

IN CASE 2012
ISN'T 2012.
tulipán
condoms

PHOTO Martin Sigal   GRAPHIC Pedro Losada   CORPORATION Ponce, Buenos Aires   COPY Antonio De Federico

PHOTO Menahem Reiss GRAPHIC Gideon Amichay, Tzur Golan, Yariv Twig, Meron Sasson COPY Sharon Refael

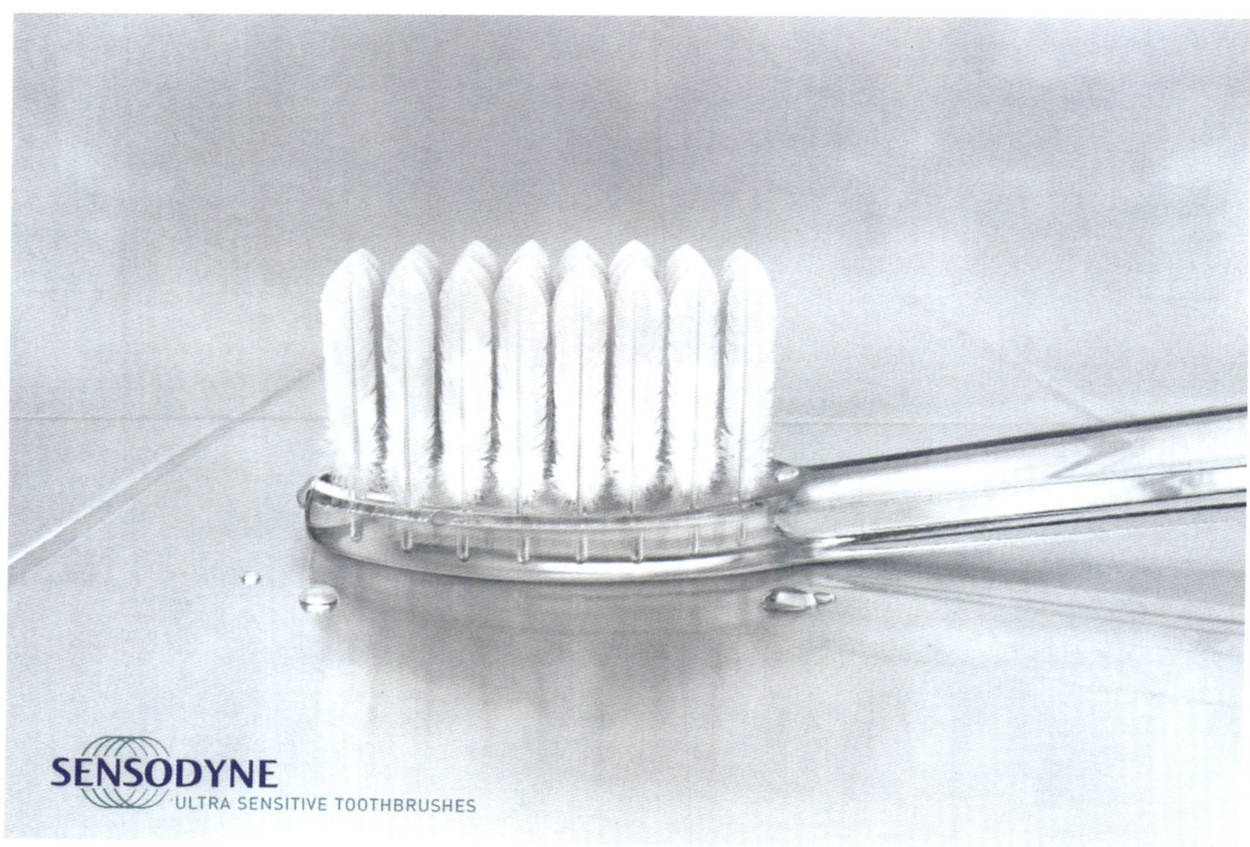

PHOTO Franz Steiner, Rainer Behrens GRAPHIC Dennis Wolf CORPORATION Grey Worldwide, Düsseldorf

GRAPHIC Alexander Nowak CORPORATION Y&R, New York COPY Feliks Richter

PHOTO Zena Holloway GRAPHIC Jay Furby CORPORATION Jay Grey, Sydney

GRAPHIC Darren Keff, Phillip Meyler CORPORATION JWT, London COPY Darren Keff, Phillip Meyler

GRAPHIC Braden Belmonte   CORPORATION Gang of One, New York   COPY Marc Stolove

PHOTO Clive Stewart   GRAPHIC Laura May Vale   CORPORATION Euro RSCG, Johannesburg   COPY Justin Wanliss

GRAPHIC Tilman Gossner CORPORATION Jung von Matt,

PHOTO Vincent Skoglund, André Thijssen

GRAPHIC Zhou Ning, Nils Andersson, Jacky Lung   CORPORATION Ogilvy, Beijing   COPY Nils Andersson

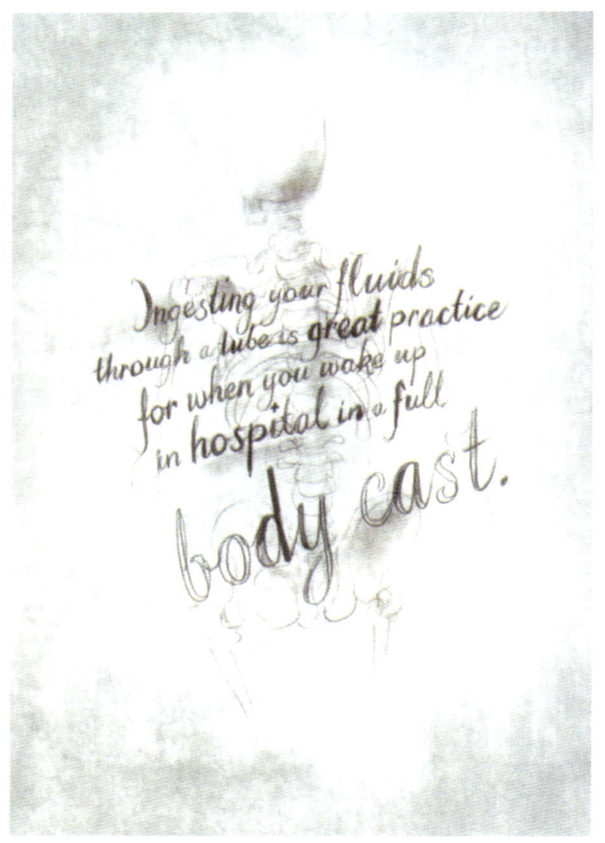

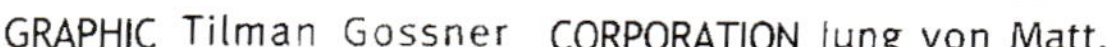

GRAPHIC Tilman Gossner CORPORATION Jung von Matt,

PHOTO Andreas Franke   GRAPHIC Joe Stitzlein   CORPORATION Nike in-house, Portland

PHOTO Nadav Kander GRAPHIC January Vernon CORPORATION Droga5, New York COPY Scott Ginsberg

PHOTO Zarella Neto GRAPHIC Fernando Reis CORPORATION Ogilvy, São Paulo COY Marcelo Padoca

PHOTO Mário Daloia   GRAPHIC Marcus Kawamura   CORPORATION Mohallem Meirelles, São Paulo   COPY Ana Carolina Reis

PHOTO Matthew Rolston   GRAPHIC Peter Barnes   CORPORATION DDB West, San Francisco   COPY Lisa Goodfriend

PHOTO Casper Sejerse CORPORATION &Co., Copenh

PHOTO Mark Seliger GRAPHIC Hoon Pin Kek, Scott McClelland CORPORATION BBH, Singapore

PHOTO Ryan McGinley GRAPHIC Wieden+Kennedy Five CORPORATION Wieden+Kennedy, Portland COPY Wieden+Kennedy Five

GRAPHIC Hendrik Frey   CORPORATION Publicis, Frankfurt am Main   COPY Konstantinos Manikas

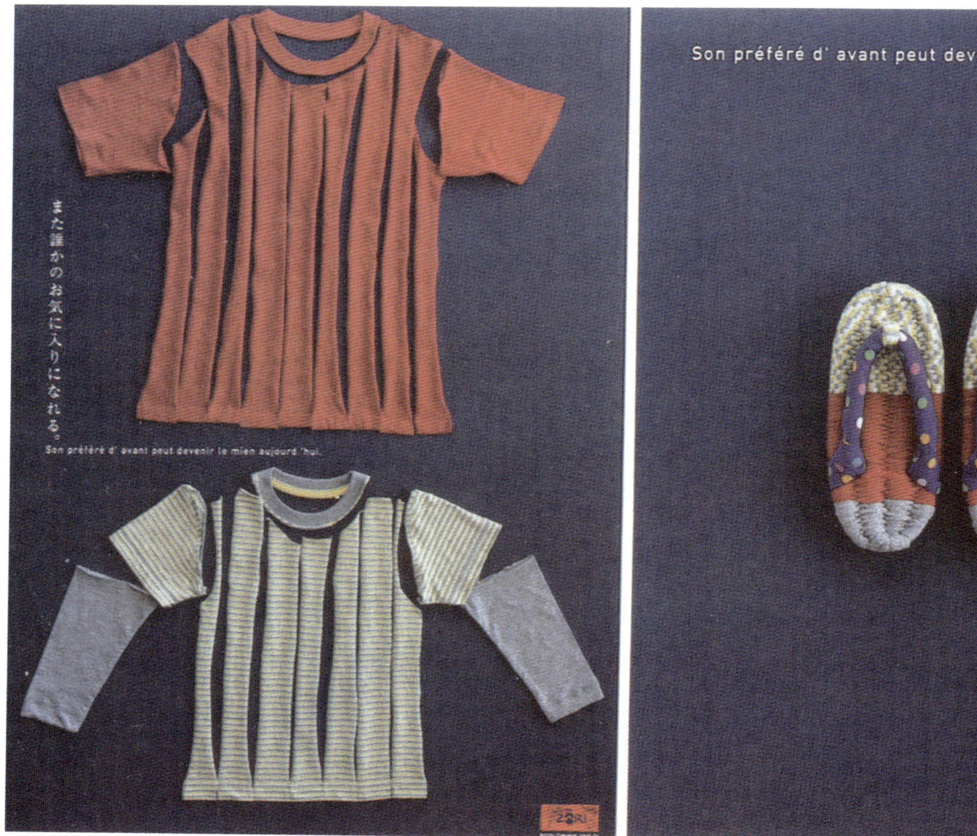

PHOTO Masahiro Sanbe   GRAPHIC Mika Niitsu   CORPORATION Monotype, Tokyo   COPY Urara Ito

PHOTO Dustin Humphrey   GRAPHIC KesselsKramer, Amsterdam   CORPORATION Y&R, Singapore   COPY Zack McDonald

PHOTO Billy Plummer   GRAPHIC Tom Hoskins   CORPORATION Iris, Sydney   COPY Jon Kelly

CORPORATION Ogilvy, Beijing   COPY Nils Andersson

PHOTO Damien Pleming   GRAPHIC Alaster Spiers, Michael   CORPORATION Draftfcb, Sydney   COPY Simon Edwards, Josh Aitken

PHOTO Mat Baker   GRAPHIC Luke Duggan   CORPORATION The Furnace, Sydney   COPY Andrew Allsop

PHOTO Varun Katyal GRAPHIC Varun Katyal   CORPORATION Contract, New Delhi COPY Varun Katyal

GRAPHIC Kazuya Tanino   CORPORATION JWT, Tokyo   COPY Kaoru Yabe

PHOTO Michael Meyersfeld   GRAPHIC Marais Janse Van Rensburg, Keshia Meyerson

PHOTO Philip Rostron   GRAPHIC Dave Marsden   CORPORATION JWT, Toronto

GRAPHIC Astrid Germanus   CORPORATION BBDO, Dusseldorf   COPY Marie-Theres Schwingeler

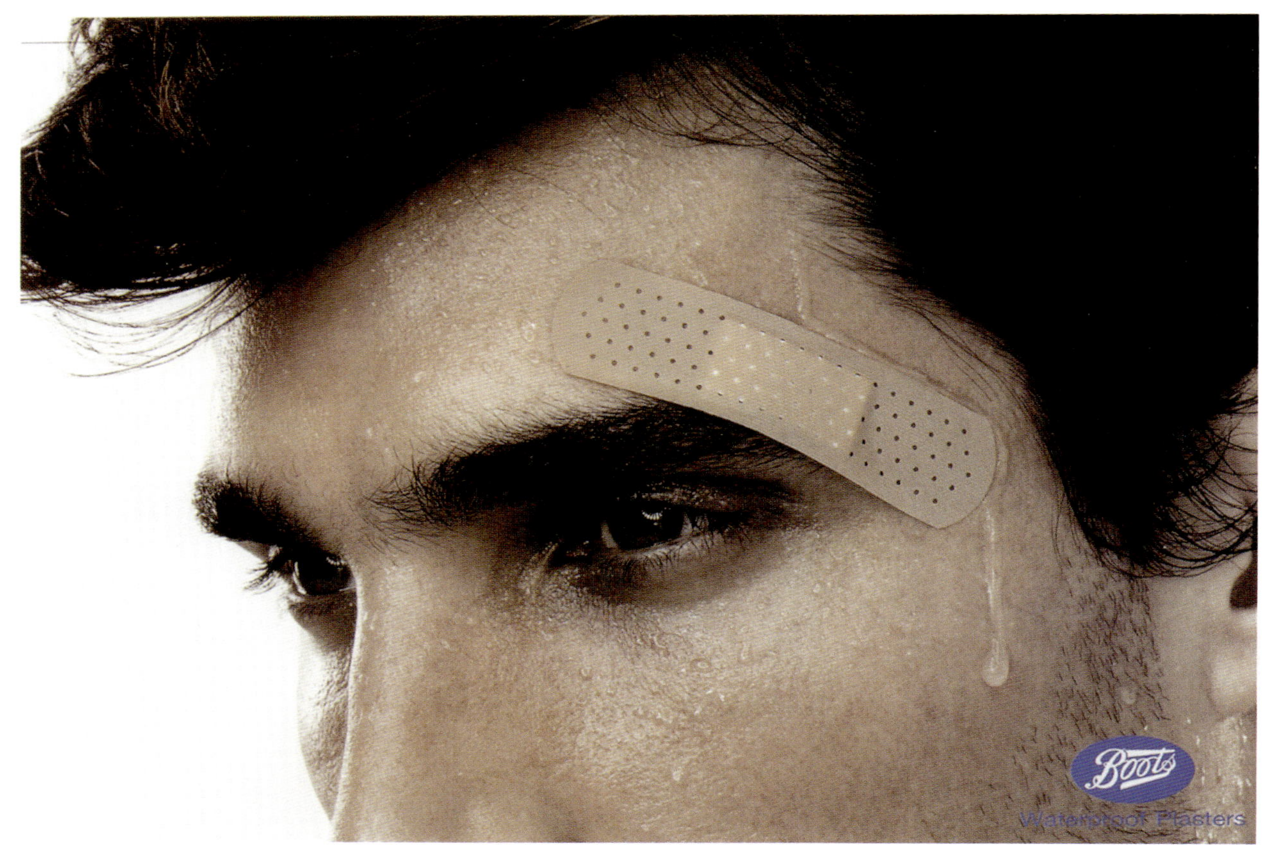

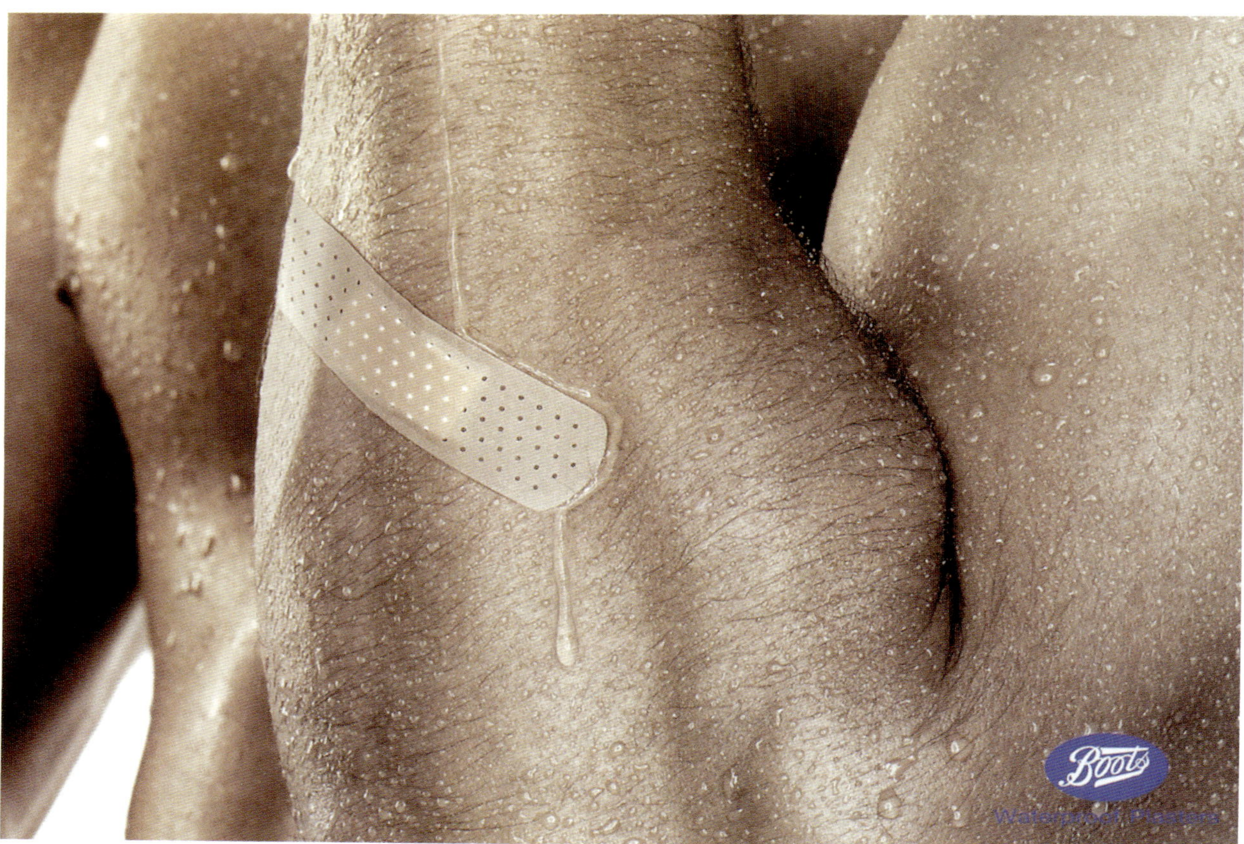

PHOTO Supachai Sriwanlapanon   GRAPHIC Boonyarit Watim   CORPORATION McCann Erickson, Bangkok

GRAPHIC Felix Tapia, Lennyn Salinas   CORPORATION JWT, Caracas

GRAPHIC Matteo Civaschi   CORPORATION McCann Erickson, Milan

PHOTO Nick Georgiou   GRAPHIC Jamie Marshall   CORPORATION Publicis, London   COPY Gary Turner

PHOTO Christophe Gilbert   GRAPHIC Didier Vandenbrandt   CORPORATION Saatchi & Saatchi, Brussels   COPY Damien Veys

GRAPHIC Orkun Önal, Can Pehlivanli   CORPORATION TBWA, İstanbul

Jockey No Panty Line

GRAPHIC Nikhil Narayanan, Joono Simon   CORPORATION DDB Mudra, Bangalore   COPY Akhilesh Bagri

PHOTO Casper Jackerhausen-Sejersen   GRAPHIC Thomas Hoffmann, Martin Storgaard   CORPORATION & Co., Copenhagen

PHOTO Hans Starck   GRAPHIC Gabriel Mattar, Johannes Hicks, Stefan Schulte   CORPORATION DDB, Berlin

PHOTO Christophe Gilbert   **GRAPHIC** Phil van Duynen

PHOTO Riccardo Bagnoli   GRAPHIC Giovanni Settesoldi   CORPORATION Grey, Paris   COPY Luissandro Del Gobbo

PHOTO Dragos Coman   GRAPHIC Holger Paasch, Mihai Coliban   CORPORATION BBDO, Moscow

PHOTO Joan Garrigosa   GRAPHIC Maged Nassar, Fadi Yaish   CORPORATION FP7, Dubai   COPY Mohamed Diaa

PHOTO Steve Hiett GRAPHIC Pierre Dupaquier, Clément Durou CORPORATION La Chose, Paris

PHOTO Riccardo Bagnoli   GRAPHIC Giovanni Settesoldi, Sebastian Burghardt

SMART MAY HAVE THE BRAINS, BUT STUPID HAS THE BALLS.
BE STUPID
DIESEL
FOR SUCCESSFUL LIVING

WASTE PAPER
STUPID CREATES.
SMART CRITIQUES.
BE STUPID
DIESEL
FOR SUCCESSFUL LIVING

PHOTO York Christoph   GRAPHIC Björn von Buchholtz, Zuzana Havelcová, Timm Holm   CORPORATION Butter, Berlin

GENERAL
COMPARTMENT

PHOTO Oliver Elliott   GRAPHIC Tim Clegg   CORPORATION Iris, London   COPY Phil Kitching

GRAPHIC Juliet Honey, Miguel Nunes   CORPORATION Lowe Bull, Johannesburg   COPY Konstant Van Huyssteen

PHOTO Gary Land GRAPHIC Martin Terhart, Daniel Carlsson, Alan von Lutzau CORPORATION 180 Amsterdam, Amsterdam

"SOMETIMES TO KNOCK DOWN A WALL YOU JUST NEED TO TOUCH IT FIRST."
FRANCESCA HALSALL
Silver World Medal
100m Freestyle
arena
WATER INSTINCT

ARENA TEAM
arena
"I KNOW THE SECRET OF TURNING WATER INTO GOLD."
ALAIN BERNARD
Olympic Champion
100m Freestyle
arena
WATER INSTINCT

In Haltern gibt es keine Industrie.
DAFÜR TORE
AM FLIESSBAND.
TU'S, HALTERN!

Haltern ist kein Kurort.
TROTZDEM GIBT'S BEI
UNS 'NE PACKUNG.
TU'S, HALTERN!

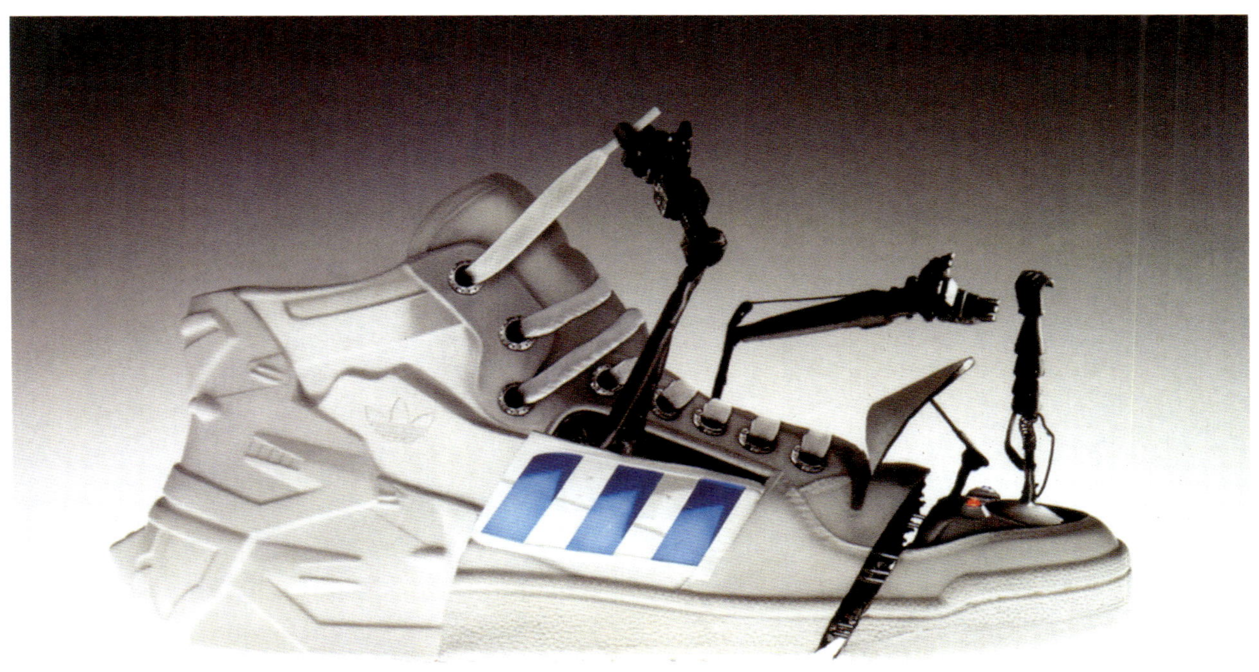

PHOTO Christopher Tovo   GRAPHIC Dimitri Kalagas   CORPORATION Lifelounge, Melbourne   COPY Chris Cork

PHOTO Christian Weber   GRAPHIC Ewoudt Boonstra   CORPORATION KesselsKramer, Amsterdam   COPY Zack McDonald